Grade 4

COUNTDOWN TO COMMON CORE

Mathematics Performance Tasks

www.mheonline.com

McGraw Hill Education

www.mheonline.com

Mc Graw Hill Education

STEM McGraw-Hill is committed to providing instructional materials in Science, Technology, Engineering, and Mathematics (STEM) that give all students a solid foundation, one that prepares them for college and careers in the 21st century.

Send all inquiries to:
McGraw-Hill Education
8787 Orion Place
Columbus, OH 43240

ISBN: 978-0-02-138214-9
MHID: 0-02-138214-X

Printed in the United States of America.

2 3 4 5 6 7 8 9 RHR 17 16 15 14

Contents

Performance Task 1: It's My Party!
Overview..5–6
Implementing the Task.......................................7
Scoring Rubric...8
Student Performance Task.....................................9
Student Recording Sheets................................10–11

Performance Task 2: Community Courtyard
Overview...12
Implementing the Task......................................13
Scoring Rubric..14
Student Performance Task....................................15
Student Recording Sheets................................16–17

Performance Task 3: End-of-Year Gift
Overview...18
Implementing the Task......................................19
Scoring Rubric..20
Student Performance Task....................................21
Student Recording Sheets................................22–23

Performance Task 4: Community Sports Complex
Overview...24–25
Implementing the Task....................................26–27
Scoring Rubric...28
Student Performance Task..................................29–30
Student Recording Sheets................................31–32

To the Teacher

This booklet contains four performance tasks for Grade 4. The purpose of a performance task is to give students experience in solving real-world situations that require critical thinking and communication of the problem-solving processes used to answer the given prompts. Each task is based on a scenario, resembling an in-depth project with multiple parts that demonstrate the mathematical understanding prescribed by the Common Core State Standards (CCSS). The tasks are exercises in rigor that require multi-step thinking and assimilation of content.

The tasks should be completed after the students have been exposed to the standards and content associated with the task. Students will choose strategies they have learned to plan and execute a solution to each step in the task. They will be required to form conclusions, make decisions, and express the rationale behind their thoughts.

After the performance task is completed you may use this as an opportunity for formative assessment as student share their results with others in the class. Students can compare their plans and results to demonstrate that different strategies may get them to the same result or to other viable results. You can also use these as group activities.

Each task is composed of four components:

Teacher Support
Includes an overview, helpful hints for students, necessary skills and tools needed to complete the task, CCSS content and Mathematical Practices imbedded in the task, as well as depth of knowledge information

Scoring Rubric
For both student self-evaluation and teacher evaluation of the work done on the performance task in one useful page that can be used as a guide of student expectations, as well as a final evaluation tool

Performance Task masters
Instructions and prompts for the various steps of the performance task

Student Recording Pages
Furnished grids and workspace for recording final results of each step in the performance task

Use this tool to give you and your students an interesting and exciting way to evaluate what they have learned this year.

Performance Task 1: It's My Party!

CCSS

Common Core State Standards
Grade 4 Mathematics

Domains and *Clusters*	OA	**Operations and Algebraic Thinking** *Use the four operations with whole numbers to solve problems. Gain familiarity with factors and multiples. Generate and analyze patterns.*
	NBT	**Number and Operations in Base Ten** *Use place value understanding and properties of operations to perform multi-digit arithmetic.*
	NF	**Number and Operations-Fractions** *Extend understanding of fraction equivalence and ordering. Build fractions from unit fractions by applying and extending previous understandings of operations on whole numbers.*
	MD	**Measurement and Data** *Solve problems involving measurement and conversion of measurements from a larger unit to a smaller unit.*
Mathematical Practices	MP1	Make sense of problems and persevere in solving them.
	MP2	Reason abstractly and quantitatively.
	MP3	Construct viable arguments and critique the reasoning of others.
	MP4	Model with mathematics.
	MP5	Use appropriate tools strategically.
	MP6	Attend to precision.
	MP7	Look for and make use of structure.
	MP8	Look for and express regularity in repeated reasoning.
Mathematics Standards	4.OA.3	Solve multistep word problems posed with whole numbers and having whole-number answers using the four operations, including problems in which remainders must be interpreted. Represent these problems using equations with a letter standing for the unknown quantity. Assess the reasonableness of answers using mental computation and estimation strategies including rounding.
	4.OA.4	Find all factor pairs for a whole number in the range 1–100. Recognize that a whole number is a multiple of each of its factors. Determine whether a given whole number in the range 1–100 is a multiple of a given one-digit number. Determine whether a given whole number in the range 1–100 is prime or composite.
	4.OA.5	Generate a number or shape pattern that follows a given rule. Identify apparent features of the pattern that were not explicit in the rule itself.
	4.NBT.5	Multiply a whole number of up to four digits by a one-digit whole number, and multiply two two-digit numbers, using strategies based on place value and the properties of operations. Illustrate and explain the calculation by using equations, rectangular arrays, and/or area models.
	4.NF.1	Explain why a fraction a/b is equivalent to a fraction $(n \times a)/(n \times b)$ by using visual fraction models, with attention to how the number and size of the parts differ even though the two fractions themselves are the same size. Use this principle to recognize and generate equivalent fractions.
	4.NF.2	Compare two fractions with different numerators and different denominators, e.g., by creating common denominators or numerators, or by comparing to a benchmark fraction such as 1/2. Recognize that comparisons are valid only when the two fractions refer to the same whole. Record the results of comparisons with symbols >, =, or <, and justify the conclusions, e.g., by using a visual fraction model.

4.NF.3 Understand a fraction *a/b* with *a* > 1 as a sum of fractions 1/*b*.

    a. Understand addition and subtraction of fractions as joining and separating parts referring to the same whole.

    c. Add and subtract mixed numbers with like denominators, e.g., by replacing each mixed number with an equivalent fraction, and/or by using properties of operations and the relationship between addition and subtraction.

    d. Solve word problems involving addition and subtraction of fractions referring to the same whole and having like denominators, e.g., by using visual fraction models and equations to represent the problem.

4.NF.4 Apply and extend previous understandings of multiplication to multiply a fraction by a whole number.

    b. Understand a multiple of *a/b* as a multiple of 1/*b* and use this understanding to multiply a fraction by a whole number.

    c. Solve word problems involving multiplication of a fraction by a whole number, e.g., by using visual fraction models and equations to represent the problem.

4.MD.2 Use the four operations to solve word problems involving distances, intervals of time, liquid volumes, masses of objects, and money, including problems involving simple fractions or decimals, and problems that require expressing measurements given in a larger unit in terms of a smaller unit. Represent measurement quantities using diagrams such as number line diagrams that feature a measurement scale.

Depth of Knowledge

Item	DOK	Key Actions
1A	2, 3	Identify patterns; Interpret; Construct
1B	2, 4	Calculate; Identify patterns; Synthesize; Apply concepts
2A	2, 4	Calculate; Arrange; Illustrate; Apply concepts
2B-2C	2	Calculate; Interpret
2D	3, 4	Draw conclusions; Apply concepts
3A	2	Calculate; Construct; Show
3B	4	Calculate; Apply concepts
3C	2, 4	Calculate; Apply concepts; Predict
3D	2, 3, 4	Compare; Draw conclusion; Cite evidence; Synthesize

Task Scenario

Students will assist with party details by interpreting data to inform guest list decisions, creating a plan for a cupcake display, and determining the amount and cost of the frozen yogurt needed for the party.

Relating to the Task

This performance task will assess students' ability to problem solve using the four operations with whole numbers and fractions. Before starting the task, you may wish to first expose them to the real-world scenario they will experience:

- Have students discuss their previous experiences with different kinds of parties (i.e. birthday, holiday, wedding, school).

- Bring in a party organizer/planner to talk about the tasks that need to be completed in order to have a successful party.

- Have students create a to-do list for a party and share their thoughts on how to host a fun and exciting party.

Implementing the Task

The following suggestions include questions to spur discussion or to prompt students who may have difficulty in progressing through the tasks on their own.

Item	Teacher Notes
1A	Give support to students who may not know how to draw a table or determine which facts to include. **What information did you include in your table?** A row for each of the years 1-5 and a row for the number of invited guests for each year. **Expected final response:** 72 guests are invited this year; Sample answers: The rule or pattern is ×3, ×5, ×7, ×9; add 16
1B	Advise students to apply concepts they know about multiples, factors, and patterns to determine which statements are true. **Expected final response:** T; F–The other factor, when multiplied by 8, is an odd number; F–The number of guests invited for the next four years would be 88, 104, 120, and 136; T
2A	**How did you represent each flavor in the visual fraction model?** Sample answers: with the beginning letters of each flavor; with the color of the flavor. **Expected final response:** 4 trays; 7 almond, 9 chocolate, 4 buttercream, 5 strawberry, 12 mint
2B	The students will need to be sure all the trays are filled with 10 cupcakes before they begin filling a new tray. When counting the empty spots they need to write this number as a fractional part of 10 since 1 tray is one whole. **Expected final response:** All of the trays will not be filled; three cupcakes could still fit on the last tray; $\frac{3}{10}$ of one tray will not be filled.
2C	Some students may have difficulty because they may write the total number of cupcakes over the total number of spaces $\left(\frac{37}{50}\right)$. They need to write their improper fraction to show the total number of cupcakes over a denominator of 10 since 1 tray is one whole. **Expected final response:** $\frac{37}{10}$ trays were filled; $\frac{10}{10} + \frac{10}{10} + \frac{10}{10} + \frac{7}{10} = \frac{37}{10}$
2D	Watch for students who say there are more chocolate cupcakes because the numerator is greater. Help them recognize when denominators are different an equivalent fraction, $\frac{6}{5}$, needs to be found. **Expected final response:** mint; there are $\frac{9}{10}$ chocolate cupcakes and $\frac{6}{5}$ or $\frac{12}{10}$ mint
3A	Encourage students to use multiplication, even though they could use repeated addition. **Did someone find this answer by using a strategy other than multiplying?** Sample answer: I added $\frac{5}{12} + \frac{5}{12} + \frac{5}{12} + \frac{5}{12} = \frac{20}{12} = 1\frac{8}{12}$ or $1\frac{2}{3}$ **Expected final response:** $1\frac{2}{3}$ quarts per table; $1\frac{2}{3}$ should be between 1 and 2.
3B	**What operation would you most likely use to answer this question?** Sample answer: multiplication **Expected final response:** 30 quarts
3C	**Why do you need 8 gallons if $30 \div 4 = 7\frac{1}{2}$ gallons?** There is a remainder of $\frac{1}{2}$, so in order to have enough yogurt you need to round up to the next whole gallon. **Expected final response:** 8 gallons; $24
3D	The students most likely will use multiplication and comparison of numbers to answer the question; however, they may have used alternate strategies. Encourage students to share these strategies. **Expected final response:** Gallons are less expensive; $30 \times \$1 = \$30$; $8 \times \$3 = \$24$; $\$24 < \$30$

Name _____

	Scoring Rubric	
Part	**What I Did**	**How I Scored**
1A	☐ I drew and extended the table correctly. **1 point** ☐ I identified the number of guests and the rule or pattern. **2 points**	____ ____ ☐/3
1B	☐ I answered _____ out of 4 true or false statements correctly. **4 points**	☐/4
2A	☐ I drew and labeled a visual fraction model. **2 points** ☐ I determined the total number of trays needed. **1 point**	____ ____ ☐/3
2B	☐ I wrote a fraction for the number of empty spots. **1 point**	____ ☐/1
2C	☐ I added the fractions and wrote the correct improper fraction. **2 points** ☐ I wrote an equation to justify my answer. **1 point**	____ ____ ☐/3
2D	☐ I correctly identified which flavor there was more of and explained my answer. **2 points**	____ ☐/2
3A	☐ I correctly calculated the number of quarts of frozen yogurt needed for each table. **1 point** ☐ I correctly labeled my number line and my answer. **2 points**	____ ____ ☐/3
3B	☐ I correctly calculated the quarts needed. **1 point**	____ ☐/1
3C	☐ I converted quarts to gallons and rounded up. **2 points** ☐ I calculated the cost of frozen yogurt in gallons. **1 point**	____ ____ ☐/3
3D	☐ I determined whether quarts or gallons was less expensive and reasoned with equations, pictures, or words. **2 points**	____ ☐/2
	Total Points	☐/25

Copyright © McGraw-Hill Education

Name _____

Performance Task 1: It's My Party!

You are planning an end-of-the-year party for school. You will interpret guest list data from past years and identify patterns, create a plan for a cupcake display, and determine the amount of frozen yogurt needed. Record your answers for each part on the Student Recording Sheet.

1. Guest List

A. This is the 5th year for the end-of-the-year party for school. The first year, 8 guests were invited. The second year, 24 guests were invited. The third year, 40 guests were invited. Last year, 56 guests were invited. Generate a table to identify a rule or pattern of invited guests for each year. Extend the pattern to determine the correct number of guests to invite to this year's party.

B. Choose whether each statement in Part 1B on your Recording Sheet is *true* or *false*.

2. Display for Cupcakes

Determine the number of trays you will need to display the cupcakes. Each tray can hold 10 cupcakes. $\frac{7}{10}$ of a tray will be almond, $\frac{9}{10}$ of a tray will be chocolate, $\frac{4}{10}$ of a tray will be buttercream, $\frac{5}{10}$ strawberry, and $\frac{6}{5}$ mint.

A. Illustrate the number of trays you need in order to display all of the cupcakes.

B. Tell what fractional part of the last tray can still be filled with cupcakes?

C. Tell what number can be represented by the filled parts of the trays. Write it as an improper fraction. Justify your answer with an equation.

D. Are there more chocolate or mint cupcakes? Explain your answer.

3. Buying Frozen Yogurt

A. If each person at a table of four is going to eat $\frac{5}{12}$ of a quart of frozen yogurt, how many quarts will you need for each table? Between what two whole numbers does this answer lie? Construct a number line to show your answer.

B. How many quarts will you need for the entire party if there are 18 tables of 4 guests?

C. How many gallons are needed. The cost of a gallon of frozen yogurt is $3. How much will you spend if you bought gallons of frozen yogurt?

D. Quart-size containers of frozen yogurt are $1 each. Would buying gallons or quarts be less expensive? Show your reasoning by using equations, pictures, and/or words.

Name _____

Student Recording Sheet

Part 1A

_____ guests are invited to this year's party.

The pattern is _____.

Part 1B

There is a pattern to the number of guests invited each year. Choose whether each statement is true or false.

True	False	
◯	◯	The number of guests invited each year is a multiple of 8.
◯	◯	Each factor of the number of guests invited each year, when multiplied by 8, is an even number.
◯	◯	If the same pattern was extended for the next four years, the number of guests invited would be 88, 96, 104, and 120.
◯	◯	The number of invited guests is always an even number.

Part 2A

This visual fraction model shows the number of trays needed to display the cupcakes.

_____ trays are needed.

Student Recording Sheet

Part 2B

Part 2C

Part 2D

There are more _____ cupcakes

because _____

_____.

Part 3A

_____ quarts

Part 3B

_____ quarts

Part 3C

_____ gallons

will cost $ _____.

Part 3D

_____

_____

Performance Task 2: Community Courtyard

CCSS

Common Core State Standards
Grade 4 Mathematics

Domains and *Clusters*	OA	**Operations and Algebraic Thinking** *Use the four operations with whole numbers to solve problems*
	NBT	**Number and Operations in Base Ten** *Use place value understanding and properties of operations to perform multi-digit arithmetic.*
	MD	**Measurement and Data** *Solve problems involving measurement and conversion of measurements from a larger unit to a smaller unit.*
Mathematical Practices	MP1	Make sense of problems and persevere in solving them.
	MP2	Reason abstractly and quantitatively.
	MP4	Model with mathematics.
	MP5	Use appropriate tools strategically.
	MP6	Attend to precision.
	MP7	Look for and make use of structure.
Mathematics Standards	4.OA.3	Solve multistep word problems posed with whole numbers and having whole-number answers using the four operations, including problems in which remainders must be interpreted. Represent these problems using equations with a letter standing for the unknown quantity. Assess the reasonableness of answers using mental computation and estimation strategies including rounding.
	4.NBT.5	Multiply a whole number of up to four digits by a one-digit whole number, and multiply two two-digit numbers, using strategies based on place value and the properties of operations. Illustrate and explain the calculation by using equations, rectangular arrays, and/or area models.
	4.MD.3	Apply the area and perimeter formulas for rectangles in real world and mathematical problems.

Depth of Knowledge

Item	DOK	Key Actions
1A	3, 4	Develop a logical argument; Apply concepts; Prove
1B	2, 3	Calculate; Interpret; Assess
2A	2, 3	Formulate; Construct; Draw
2B	2, 4	Create; Design; Construct; Label; Calculate
3A	2, 4	Estimate; Apply concepts; Calculate
3B	2, 4	Estimate; Organize; Apply concepts;
3C	2, 3, 4	Estimate; Tabulate; Cite Evidence; Prove

Task Scenario

Students create a layout and a budget for constructing a community courtyard within given size and space restrictions.

Relating to the Task

This performance task will assess students' ability to problem solve with measurement in the context of area/space. Before giving students this task, you may wish to first expose them to the real-world scenario they will experience:

- Bring in a local engineer or landscaper to share his or her expertise with the class.

- Show examples of courtyards in communities or in famous locales.

- Bring in flower boxes with soil and flowers. Have students measure the box and think about how many plants might fit in the box.

- Have students create a simple spreadsheet to keep track of a budget.

Implementing the Task

The following suggestions include questions to spur discussion or to prompt students who may have difficulty in progressing through the task on their own.

Item	Teacher Notes
1A	Encourage students to use the formula to find the perimeter and area. If students find the perimeter or area another way, have them explain their strategy. ***How might you find the area if you do not use the formula? Explain.*** Sample answer: I drew equal sized squares in the rectangle and counted the squares. **Expected final response:** 8 feet and 4 feet; 8 + 4 + 8 + 4 = 24 feet; 8 × 4 = 32 square feet
1B	***How did you determine the area of the Common Space?*** Sample answer: I measured the length and width by counting the squares and then used the area formula to find the area. ***How did you evaluate each statement to see if it was true or false?*** Sample answer: I compared each statement to the information that was provided. Then I used what I know about area to determine if the statement was true. **Expected final response:** 15 feet and 15 feet; T; F–The area is 225 square feet; F– I do have enough information to find the area; T; T
2A	Make sure students are following the size and space requirements outlined in this part of the task. Advise cutting out a table template and placing the tables along the perimeter of the Common Space. ***How do you know the tables are 3 square feet by 3 square feet? Explain.*** Sample answer: I counted three squares down and three squares across for each. **Expected final response:** 9 tables
2B	Students should create their own layout. There may be many different designs. Advise them to recall how they previously found the area. **Expected final response:** See students' work for flower beds.
3A	In all of Part 3 help students recall the importance of estimating in order to check their answers for reasonableness. ***How can you find the total area of all the flower beds?*** Sample answer: I can find the area of each flower bed and then add to find the total area. ***How can you find the total cost?*** Sample answer: Since each bag of mulch covers 3 square feet I have to divide the total area of all the flower beds by 3 to find how many bags I will need. Then, I can multiply the number of bags by the cost per bag. **Expected final response:** See students' work. Number and size of new gardens will determine this answer and answers will vary.
3B	***How can you find the cost for all of the tables?*** Sample answer: I can multiply the cost of each table, $60, by the number of tables I calculated. **Expected final response:** $60 × 9 = $540
3C	***Describe how you could use addition and multiplication to find the total cost of the flowers.*** I could add $8 + $8 + $8 + $8 + $8 + $8 + $8 + $8 + $8 + $8 or find 10 × $8. **Expected final response:** $80 for the flowers; The total cost of the project will vary due to the number of bags of mulch needed for the flower beds.

Scoring Rubric		
Part	**What I Did**	**How I Scored**
1A	☐ I used the area and perimeter formulas to calculate the area of the flower bed. **1 point** _____ ☐ I calculated the area and perimeter correctly. **2 points** _____	☐/3
1B	☐ I answered _____ out of 5 true and false statements correctly. **5 points**	☐/5
2A	☐ I drew tables that were 3 feet by 3 feet on the map. **1 point** _____ ☐ I drew the tables with at least a 3 feet distance from each other. **1 point** _____ ☐ I found the greatest number of tables that would fit in the space. **1 point** _____	☐/3
2B	☐ I drew flower beds on the map and labeled the length and width of each. **1 point** _____ ☐ I correctly calculated and then recorded the correct length, width, and area of each flower bed. **1 point** _____	☐/2
3A	☐ I correctly calculated the total area of all the flower beds. **1 point** _____ ☐ I correctly calculated and then recorded in the chart the number of bags of mulch needed. **1 point** _____ ☐ I estimated, and then correctly calculated and recorded in the chart the total cost of the mulch. **1 point** _____	☐/3
3B	☐ I calculated the total cost of the tables. **1 point** _____ ☐ I correctly calculated and then recorded the total cost of the tables. **1 point** _____	☐/2
3C	☐ I correctly calculated and then recorded in the chart the total cost of the flats of flowers. **1 point** _____ ☐ I correctly tabulated the total cost for the project. **1 point** _____	☐/2
	Total Points	☐/20

Name _____

Performance Task 2: Community Courtyard

Your class is helping the director of a community center update the courtyard. You will develop an action plan to implement the landscaping improvements needed. You will design a layout for the courtyard and create a budget for expenses. Record your answers for each part on the Student Recording Sheet.

1. Working with Measurements

A. Use the Community Center Courtyard Map to find the length and width of the flower bed. Calculate its perimeter and area. Explain your reasoning to justify your conclusion.

B. What is the length and width of the the Common Space on the Community Center Courtyard Map? Calculate its area and choose whether each statement on the recording sheet is *true* or *false*.

2. Constructing a Plan

A. The Community Center purchased tables that are 3 feet by 3 feet for the Common Space. Formulate a plan to find the greatest number of tables that will fit while keeping at least 3 feet of distance between each table. Draw the tables on the map. Record the total number of tables in Part 2A.

B. The additional space outside the Common Space will be used for more flower beds of different lengths and widths. Draw each new flower bed on the the map. Then label each with a number, beginning with the number 2. Find each length, width, and area and record each in the chart.

3. Create a Budget

Calculate the budget for the mulch, flowers, and tables. For each part, record your findings in the chart. Show your work!

A. Find the total area of all the flower beds. Each bag of mulch costs $3 and covers 3 square feet. Find the number of bags of mulch that are needed. Estimate and calculate the total cost of the mulch.

B. Refer to your work for Part A, then calculate the total cost of the tables if they are $60 each.

C. The Community Center has decided that they want 10 flats of flowers to plant. Each flat costs $8. Tabulate the total cost of the flowers. Finally, calculate the total cost of the whole project. Present the project's design and budget to prove your findings.

Name _____

Student Recording Sheet

Community Center Courtyard Map

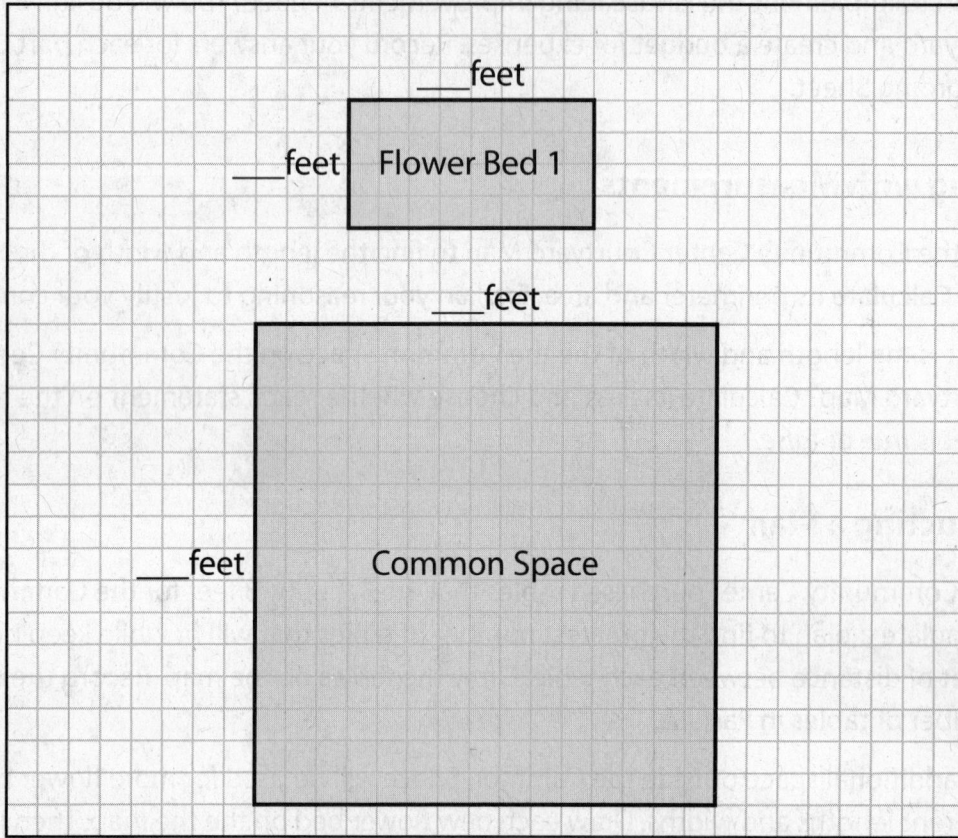

____feet

____feet | Flower Bed 1

____feet

____feet | Common Space

1 square = 1 sq ft

Part 1A

The perimeter of the flower bed is
_____ feet.

The area of the flower bed is
_____ square feet.

My reasoning proves this is correct because

_____

_____

_____

_____.

Part 1B

True	False	
○	○	This area is measured in square feet.
○	○	The area is 30 square feet.
○	○	I don't have enough information to find the area.
○	○	The area is 225 square feet.
○	○	This area is larger than the flower bed.

Student Recording Sheet

Part 2A

_____ tables

Part 2B

Flower Bed #	Width (ft)	Length (ft)	Area (sq ft)
1			

Part 3A–3C

Item	Number of Items	Cost for Each	Estimate	Actual Total
Mulch		$3 each bag		
Tables		$60 each table		
Flowers	10 flats	$8 each flat		
			Total cost	$_____

Performance Task 3: End-of-Year Gift

Common Core State Standards
Grade 4 Mathematics

Domains and Clusters	**MD**	**Measurement and Data**	*Geometric measurement: understand concepts of angle and measure angles.*
	G	**Geometry**	*Draw and identify lines and angles, and classify shapes by properties of their lines and angles.*

Mathematical Practices	**MP1**	Make sense of problems and persevere in solving them.
	MP2	Reason abstractly and quantitatively.
	MP3	Construct viable arguments and critique the reasoning of others.
	MP4	Model with mathematics.
	MP5	Use appropriate tools strategically.
	MP6	Attend to precision.
	MP7	Look for and make use of structure.

Mathematics Standards	**4.MD.5**	Recognize angles as geometric shapes that are formed wherever two rays share a common endpoint, and understand concepts of angle measurement:
	4.MD.6	Measure angles in whole-number degrees using a protractor. Sketch angles of specified measure.
	4.MD.7	Recognize angle measure as additive. When an angle is decomposed into non-overlapping parts, the angle measure of the whole is the sum of the angle measures of the parts. Solve addition and subtraction problems to find unknown angles on a diagram in real world and mathematical problems, e.g., by using an equation with a symbol for the unknown angle measure.
	4.G.1	Draw points, lines, line segments, rays, angles (right, acute, obtuse), and perpendicular and parallel lines. Identify these in two-dimensional figures.
	4.G.2	Classify two-dimensional figures based on the presence or absence of parallel or perpendicular lines, or the presence or absence of angles of a specified size. Recognize right triangles as a category, and identify right triangles.
	4.G.3	Recognize a line of symmetry for a two-dimensional figure as a line across the figure such that the figure can be folded along the line into matching parts. Identify line-symmetric figures and draw lines of symmetry.

Depth of Knowledge

Item	DOK	Key Actions
1A	2	Measure; Summarize
1B	2, 3	Recognize; Classify; Cite Evidence
1C	2, 4	Measure; Identify; Construct; Apply Concepts
2A	3, 4	Draw conclusions; Infer; Develop a logical argument; Apply Concepts
2B	2, 3	Draw; Classify; Interpret; Construct
2C	2	Identify, Label; Categorize
3A	3, 4	Draw; Prove; Develop a logical argument
3B	3, 4	Draw; Explain phenomena in terms of concepts; Analyze

Task Scenario

Students design quilt squares based upon angle measurements, attributes of two-dimensional figures, and line of symmetry.

Relating to the Task

This performance task will assess students' ability to problem solve using measurement and geometry in the context of angles and figures. Before giving students this task, you may wish to first expose them to the real-world scenario they will experience:

- Have some quilters come in and share how they use geometry in their work.
- Have students create quilt squares using paper or fabric.
- Display quilts and discuss the lines and angles of the two-dimensional figures.

Implementing the Task

The following suggestions include questions to spur discussion or to prompt students who may have difficulty in progressing through the tasks on their own.

Item	Teacher Notes
1A	**Summarize and explain how you measured the angle.** Sample answer: Using a protractor, I lined up its center with the shared endpoint of the angle keeping its straight edge along one ray, pointing at zero. Then I found the tic mark on the protractor that lined up with the second ray of the angle. **Expected final response:** 105°
1B	**What evidence supports your angle classification?** Sample answer: I know that a right angle measures 90°, an acute angle measures greater than 0° and less than 90°, and an obtuse angle measures greater than 90° and less than 180°. **Expected final response:** acute angle
1C	**Explain what you were thinking in order to complete this problem.** Sample answer: I know that I can add the measure of each individual angle to get the measure of the combined angle, so I found 105° + 20° = 125°. **Expected final response:** The students' angles should measure 125°; 125°
2A	**How did you determine which statements were true and which were false?** Sample answer: I looked at each figure and statement, and then asked myself "Does that figure fit the statement?" **Expected final response:** True; True; False; True; False
2B	**Will everyone's figures be the same?** Yes **What can you infer from this?** Sample answer: A figure's attributes help classify it into certain categories. **Expected final response:** square; rhombus; rectangle; trapezoid
2C	**Would you expect that everyone's response to be the same for each attribute?** **Explain your reasoning.** No; Sample answer: A figure may have more than one attribute. **Expected final response:** Students' responses will vary.
3A	**Expected final response:** No, he is not correct because Figure A and E do not have line symmetry. There is not a way to fold either in half so each of the two sides match.
3B	**Expected final response:** Figures B, C, and D have only one line of symmetry. Figure F has five lines of symmetry.

Scoring Rubric		
Part	**What I Did**	**How I Scored**
1A	☐ I measured the angle correctly. **1 point**	___ ☐/1
1B	☐ I classified the angle correctly. **1 point**	___ ☐/1
1C	☐ I drew the two angles as one correctly. **1 point** ☐ I correctly identified the measurement of the new angle. **1 point**	___ ___ ☐/2
2A	☐ I correctly identified _____ out of 5 statements. **5 points**	___ ☐/5
2B	☐ I correctly drew and classified the four figures with the given attributes. **4 points**	___ ☐/4
2C	☐ I correctly identified and labeled a point, line, line segment, ray, right angle, acute angle, obtuse angle, perpendicular lines, and parallel lines. **9 points**	___ ☐/9
3A	☐ I correctly identified the figures that have line symmetry. **4 points** ☐ I correctly placed each line of symmetry. **4 points**	___ ___ ☐/8
3B	☐ I correctly identified the figure that has more than one line of symmetry. **1 point** ☐ I correctly identified the additional lines of symmetry. **4 points**	___ ___ ☐/5
	Total Points	☐/35

Performance Task 3: End-of-Year Gift

Your class will be designing and constructing a quilt for your teacher's end-of-year gift. You will measure and draw angles, classify figures, and determine symmetry of the figures in the quilt. Record your answers for each part on the Student Recording Sheet.

1. Quilt Angles

A. Use a protractor to measure the angle Erin drew for her triangular quilt piece.

B. Would you classify the angle Brad drew, as right, acute, or obtuse?

C. Erin and Brad put their angles together so the angles shared one ray. Draw the new angle. What is the measure of the new angle?

2. Two-Dimensional Figures

A. The students will use a variety of figures and angles in the quilt. Nathan drew 8 figures using different lines and angles and labeled them **a–h** as shown on page 22. Choose true or false for each.

B. The following attributes describe four figures on four quilt squares:
- Figure 1: four right angles and four sides of equal length
- Figure 2: no right angles, four sides of equal length, and parallel sides
- Figure 3: four right angles, four sides, opposite sides equal in length
- Figure 4: four sides, and only one pair of parallel sides

Draw and classify each of the four figures.

C. Two figures from Emanuel's quilt square are shown in Part 2C. From the list, identify each attribute by tracing it on one of the two figures. Label each with its corresponding letter.

3. Showing Symmetry

A. Landon noticed that some of the figures in quilt squares **A-F** were symmetrical. He said that figures **A**, **B**, **C**, **D**, and **F** have line symmetry. Is he correct? Explain. Draw one line of symmetry through each figure that has line symmetry to justify your answer.

B. Which figure has more than one line of symmetry? Draw the additional lines of symmetry.

Name _____

Student Recording Sheet

Part 1A

Erin's angle

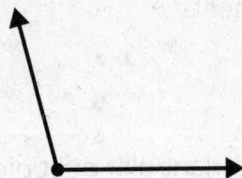

The angle measures _____.

Part 1B

Brad's angle

This is a(n) _____ angle.

Part 1C

This is Erin and Brad's new angle.

The new angle measures _____.

Part 2A

a   b   c   d

e   f   g   h

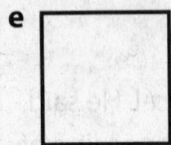

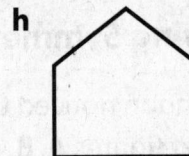

TRUE	FALSE	Statement
◯	◯	Figures **a, b, d, e, g** and **h** have at least one pair of parallel lines.
◯	◯	Figures **b, e,** and **h** have perpendicular lines.
◯	◯	Figures **b** and **e** are the ONLY figures with right angles.
◯	◯	Figures **d** and **f** have at least one obtuse angle.
◯	◯	Figures **c** and **f** have ONLY acute angles.

Grade 4 • Performance Task 3: End-of-Year Gift

Student Recording Sheet

Part 2B

Figure 1	Figure 2	Figure 3	Figure 4

Part 2C

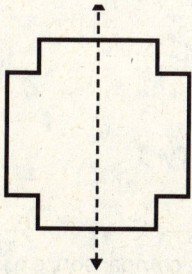 

a point
b line
c line segment
d ray
e right angle
f acute angle
g obtuse angle
h perpendicular lines
i parallel lines

Parts 3A and 3B

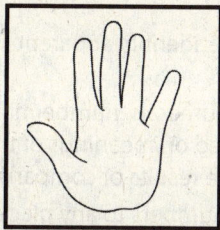

A

B

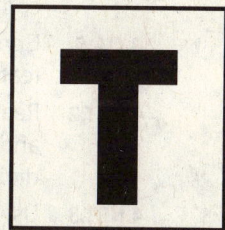

C

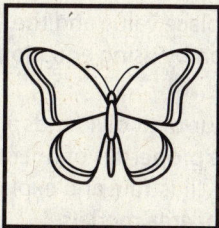

D

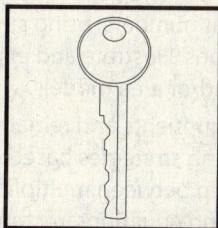

E

F

Part 3A

Part 3B

Figure _____ has more than one line of symmetry.

Performance Task 4: Community Sports Complex

Common Core State Standards
Grade 4 Mathematics

Domains and *Clusters*	**OA**	**Operations and Algebraic Thinking** *Use the four operations with whole numbers to solve problems.*
	NBT	**Number and Operations in Base Ten** *Generalize place value understanding for multi-digit whole numbers. Use place value understanding and properties of operations to perform multi-digit arithmetic.*
Mathematical Practices	**MP1**	Make sense of problems and persevere in solving them.
	MP2	Reason abstractly and quantitatively.
	MP3	Construct viable arguments and critique the reasoning of others.
	MP4	Model with mathematics.
	MP5	Use appropriate tools strategically.
	MP6	Attend to precision.
	MP7	Look for and make use of structure.
	MP8	Look for and express regularity in repeated reasoning.
Mathematics Standards	**4.OA.2**	Multiply or divide to solve word problems involving multiplicative comparison, e.g., by using drawings and equations with a symbol for the unknown number to represent the problem, distinguishing multiplicative comparison from additive comparison.
	4.OA.3	Solve multistep word problems posed with whole numbers and having whole-number answers using the four operations, including problems in which remainders must be interpreted. Represent these problems using equations with a letter standing for the unknown quantity. Assess the reasonableness of answers using mental computation and estimation strategies including rounding.
	4.OA.5	Generate a number or shape pattern that follows a given rule. Identify apparent features of the pattern that were not explicit in the rule itself.
	4.NBT.2	Read and write multi-digit whole numbers using base-ten numerals, number names, and expanded form. Compare two multi-digit numbers based on meanings of the digits in each place, using >, =, and < symbols to record the results of comparisons.
	4.NBT.3	Use place value understanding to round multi-digit whole numbers to any place.
	4.NBT.4	Fluently add and subtract multi-digit whole numbers using the standard algorithm.
	4.NBT.5	Multiply a whole number of up to four digits by a one-digit whole number, and multiply two two-digit numbers, using strategies based on place value and the properties of operations. Illustrate and explain the calculation by using equations, rectangular arrays, and/or area models.
	4.NBT.6	Find whole-number quotients and remainders with up to four-digit dividends and one-digit divisors, using strategies based on place value, the properties of operations, and/or the relationship between multiplication and division. Illustrate and explain the calculation by using equations, rectangular arrays, and/or area models.

Depth of Knowledge

Item	DOK	Key Actions
1A, 1B	2, 3	Compare; Summarize
1C	2, 3	Estimate; Compute; Infer; Draw conclusions
1D	2, 3, 4	Compare; Interpret; Identify patterns; Analyze
1E	3	Use concepts to solve non-routine problems; Cite evidence
2A	4	Calculate; Apply concepts
2B, 2C	3	Illustrate; Identify; Explain in terms of concepts
2D	4	Calculate; Prove
2E	2, 4	Calculate; Estimate; Prove
2F	3	Calculate; Draw conclusions

Task Scenario

Students assist a land developer in interpreting data gathered on three sport complexes. Rounding and comparing the data, and making calculations to find totals and differences will help inform the decisions that need to be made for a new sports complex.

Relating to the Task

This performance task will assess students' ability to problem solve by generalizing place value understanding of multi-digit whole numbers and performing multi-digit arithmetic. Before giving students this task, you may wish to first expose them to the real-world scenario they will experience:

- Have students discuss their previous experiences in sports and where they play sports.

- Bring in a recreational department manager and/or land developer to share what sports are played in the parks around the students' community.

- The students can research complexes in surrounding areas by displaying pictures and discussing the venues at the complexes.

Implementing the Task

The following suggestions include questions to spur discussion or to prompt students who may have difficulty in progressing through the task on their own.

Item	Teacher Notes
1A	***How did you determine which symbol to use?*** Sample answer: I compared the numbers in the greatest place value position, and since they were the same, I compared the thousands. Having the same number of thousands, I looked at the value of the hundreds. I know that 0 hundreds is less than 5 hundreds; therefore, I decided to use the *less than* symbol to show that Springfield has fewer seats than Ashbury. **Expected final response:** 51,092 seats < 51,566 seats ***Is it possible to have used the > symbol? Explain.*** Yes; I could have written the greater number first. 51,566 seats > 51, 092 seats
1B	***What do you need to do first to solve this problem?*** Sample answer: Find the total of the 3 events for each complex. Once I have the total, I can compare. ***How can you tell which complex has more combined seats of the three events without finding a total?*** Sample answer: The two complexes have similar totals for two events, but Ashbury has more than 12 thousand in the third event while Newton does not even have 2 thousand. **Expected final response:** 19,857 seats < 30,010 seats
1C	Be sure students remember to find an estimated total number of seats for each event. They need to estimate the number of seats at the event at each complex and then add to find the total. **Expected final response:** <table><tr><th>Event</th><th>Estimated Seat</th><th>Actual Seats</th></tr><tr><td>Baseball</td><td>99,000</td><td>99,472</td></tr><tr><td>Softball</td><td>105,000</td><td>104,643</td></tr><tr><td>Football</td><td>197,000</td><td>196,585</td></tr><tr><td>Soccer</td><td>65,000</td><td>64,693</td></tr><tr><td>Basketball</td><td>49,000</td><td>49,512</td></tr><tr><td>Swimming</td><td>28,000</td><td>27,117</td></tr></table> **Sample answer:** If funds are limited the developer may only build a sports complex for the top 3 events.
1D	**Expected final response:** The first four event's ten thousands position is 10,000 less than the event before it. Sample answer: Each event has fewer seats than all the events listed above it. ***Does that mean, "Each number is 10,000 less than the one before it?"*** No; Sample answer: Since all the other digits are not the same as the one before it you cannot make that statement.
1E	Ask students to share their strategies for solving for the number of levels. **Expected final response:** 3 levels with 627 seats on each level or 9 levels with 209 seats on each level. Sample answer: I divided the total number of seats by a single digit number until I got a quotient with no remainder.

Copyright © McGraw-Hill Education

Item	Teacher Notes
2A	***How do you know what operation to use when completing the table?*** Sample answer: If I know the number of fields/courts and the number of seats per field/court, then I have to multiply. I have to do the opposite (divide), if I know the total number of seats and the number of fields/courts to find the number of seats per field/court. **Expected final response:** 870, 900; 1,335, 400; 2,131, 2,100; 5,850, 1,200; 2,492, 2,500
2B	Students may find the answers using different strategies. Reminder: Students are not required to know the standard division algorithm until Grade 5. **Expected final response:** I used place value to find the quotient of each place value position. $3,480 \div 4 = 870$; $3480 \div 4 = (3000 \div 4) + (400 \div 4) + (80 \div 4) = 750 + 100 + 20 = 870$ Have other students explain the strategies they used that may have involved base-ten blocks, properties of operations, or the relationship between multiplication and division.
2C	Students may find the answers using different strategies. Reminder: Students are not required to know the standard division algorithm until Grade 5. **Expected final response:** Sample answer: I multiplied the number of fields by the number of seats at each field; (An illustration may be an area model that illustrates the partial products strategy); $1,170 \times 5 = 5,850$ Have other students explain the strategies they used that may have involved base-ten blocks, properties of operations, or the relationship between multiplication and division.
2D	Some students may not know whether to add or subtract. ***Explain what you were thinking when you solved this problem?*** Sample answer: I read the question and it states "how many more…than". I know when you are asking *how many more* you subtract. Encourage students to line up the place-value positions carefully and to focus on the place value when regrouping. **Expected final response:** $3,480 - 1,335 = 2,145$; $2,000 + 100 + 40 + 5$; two thousand, one hundred forty-five
2E	Some students may not know whether to add or subtract. ***How do you know what operation to use in this problem?*** Sample answer: I read the question and it states *together*, which indicates that I am making one group, adding. Encourage students to line up the place-value positions carefully and to focus on the place value when regrouping. **Expected final response:** $4,984 + 2,612 = 7,596$; $5,000 + 3,000 = 8,000$ which is close to the exact answer. So my answer is reasonable.
2F	**Expected final response:** two times more; $2 \times \underline{\quad} = 4$; Sample answer: twice as many baseball games can be going on at the same time as football games

Copyright © McGraw-Hill Education

Name _____

	Scoring Rubric		
Part	**What I Did**	**How I Scored**	
1A	☐ I used the correct symbol to show how the numbers are related to each other. **1 point**	___	☐/1
1B	☐ I computed correctly and compared the numbers using the correct symbol. **2 points**	___	☐/2
1C	☐ I correctly *estimated* and calculated the number of seats in _____ of 6 events. **12 points** ☐ I correctly ordered the popularity of each event and gave a satisfactory explanation. **2 points**	___	☐/14
1D	☐ I gave a reasonable explanation of the pattern I recognized and its meaning. **2 points**	___	☐/2
1E	☐ I correctly identified the number of levels, seats on each level, and explained my strategy. **3 points**	___	☐/3
2A	☐ I correctly found each missing value. **10 points**	___	☐/10
2B	☐ I satisfactorily illustrated or explained my strategy with equations or models. **1 point**	___	☐/1
2C	☐ I satisfactorily illustrated or explained my strategy with equations or models. **1 point**	___	☐/1
2D	☐ I used the algorithm and calculated correctly. **2 points** ☐ I correctly recorded my solution in expanded form and word form. **2 points**	___	☐/4
2E	☐ I used the algorithm, calculated correctly, and used an estimation strategy. **3 points**	___	☐/3
2F	☐ I correctly calculated my solution and made a satisfactory conclusion. **2 points**	___	☐/2
		Total Points	☐/43

Performance Task 4: Community Sports Complex

You are assisting a land developer to gather ideas for building a new sports complex. You have collected and compiled data about several sports complexes. This data will be used to make comparisons between the complexes and their events and to help generate suggestions for the new sports complex. Record your responses for each part on the Student Recording Sheet. Remember to show your work.

1. Event Seating at Three Sports Complexes

The table represents the data collected about sports complexes in surrounding communities.

Total Number of Seats at Each Complex			
Event	Springfield Complex	Newton Complex	Ashbury Complex
Baseball	63,480	32,617	3,375
Softball	51,092	1,985	51,566
Football	42,651	51,820	102,114
Soccer	35,853	14,716	14,124
Basketball	42,984	3,260	3,268
Swimming	12,618	1,881	12,618

A. Compare the number of seats at the Springfield and the Ashbury softball events. Use the correct symbol ($<$, $>$, or $=$) to summarize the results of the comparison.

B. Compare the combined number of seats at the soccer, basketball, and swimming events in Newton to the same events in Ashbury. Use the correct mathematical symbol ($<$, $>$, or $=$) to summarize the results of the comparison.

C. Find an estimated total number of seats for each event by estimating to the nearest thousand. Then calculate the actual number of seats for each event. Which event seems to be the most popular? Least popular? Order the events from most popular to least popular. Explain what this might mean to a company that is considering building a new sports complex.

D. Look at the seat totals from the first four Springfield events. What pattern do you recognize and what does that pattern tell you?

E. The number of levels in the Newton swimming event is less than 10. Each level has an equal number of seats. How many levels does the event have? How many seats are on each level? Explain your strategy for finding the number of levels.

2. Event Seating at Your Complex

The information you tabulated about the other sports complexes has helped the developer determine which events to include in the new complex and the number of seats for each.

A. The developer makes a table to represent the number of seats and fields or courts each event will have in the new complex. There will be an equal number of seats at each of the fields or courts for each event. Find the missing values and complete the table on page 32.

B. Identify your strategy for finding the number of seats at each field in the baseball event. Illustrate and explain the calculation by using equations or models.

C. Identify your strategy for finding the total number of seats at the five soccer fields. Illustrate and explain your calculation with an equation or model.

D. How many more seats are there at the baseball event than the softball event? Show your work. Record your answer in expanded form and word form.

E. How many seats are there at the swimming and basketball events together? Use the standard algorithm to solve. Justify the reasonableness of your response by using an estimation strategy.

F. The number of fields that will be at the baseball event is how many times more than the number of fields at the football event? What conclusion can you draw from this information?

Name _____

Student Recording Sheet

Part 1A

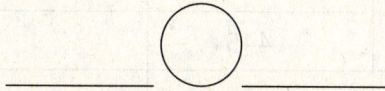

_____ _____

Part 1B

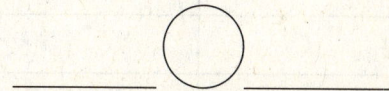

_____ _____

Part 1C

Event	Estimated Total	Actual Total
Baseball		
Softball		
Football		
Soccer		
Basketball		
Swimming		

Popularity	Event
1 (Most)	
2	
3	
4	
5	
6 (Least)	

To a company that is considering building a new sports complex, this data may tell them

_____

_____

Part 1D

_____

_____

Part 1E

The Newton swimming event has _____ levels. There are _____ seats on each level.

_____

_____

Student Recording Sheet

Part 2A

Fill in the missing values.

Event	Number of Seats	Fields/Courts	Seats per Field/Court	Round to the Nearest 100
Baseball	3,480	4		
Softball		3	445	
Football	4,262	2		
Soccer		5	1,170	
Basketball	4,984	2		
Swimming	2,612	1	2,612	2,600

Part 2B

_____

_____

_____

Part 2C

_____

_____

_____

Part 2D

_____ seats

Expanded form

_____

_____

Word form

_____

_____

Part 2E

_____ seats

_____

_____

_____

Part 2F

_____ times more

I conclude that _____

_____